Putting The Pieces Together

David Jacobson

To Glenda,
My partner for the past sixty-seven years

Table of Contents

Introduction

June 21, 1937, was the first time I breathed the air of the world. Eighty-eight years later, that air still fills my lungs. As I look back on those many years, I recall many events, especially the years of parental alcoholism repeatedly destroying my childhood, the institutions I was forced to live in, and the times I thought I would give it all up. But I survived and, in time, prospered. It occurred to me that some would associate with these events and the stories and find them interesting and perhaps beneficial. For me, the pieces detailed in this book allow me to see the complete picture of my life, the bad, the good, and the ugly.

Before I can put the pieces together, I must first gather them up. This will take us on a journey starting in San Francisco, California, moving to Denver, Colorado, Albuquerque, New Mexico, to Bartlesville, Oklahoma, then on to Sun City, Texas, Oklahoma City, Oklahoma, and back to Bartlesville, where I currently reside. We will see that my life in San Francisco was the worst, and my life in Sun City, Texas, the best.

I won't go through the table of contents here. Just look at it to see what this book is all about. I did not have a normal childhood. This story is wrapped around that not-so-normal childhood. Put the pieces together and you get me, the abused child, and the successful adult. Although I do not discuss my faults in much detail, they, too, are part of me.

Alcoholism, my parents' alcoholism, controlled the first nineteen years of my life and, to some degree, the years beyond. This book will tell the story of each piece, explain the events that took place in those years, and show how I survived them to achieve a good life. In fact, I not only survived, but I prospered. I admit that there were times when I was sorry to be alive, times when I couldn't stand the pain anymore. But with the help of others, I survived. And I survived and found my way to many joyous years. Sometimes it takes friends and, sometimes, even strangers to help you along. Sometimes it's just your struggle to be a good person that leads to a good life.

My primary goal in this book is to show that success is possible despite the sometimes seemingly unsolvable problems one may face. If you don't fight for it, success is difficult to achieve. If you are faced with such a challenge, you can overcome the difficulty.

The mood at the beginning of this book is quite somber, but it will change for the better. My adventures in the Air Force, marriage, schooling, an exciting new profession, children, computers, retirement, art, RVing, and the good life in Texas will reveal that change.

Here and there, I will tell some short stories, some good, some bad, some funny, some sad.

Now, let us start gathering the pieces of my life and put them all together.

Let us see if and how your life relates to mine.

About the Art

The artwork appearing in this book is original digital art (computer art) created by the author over many years. The author has selected most of them because they relate in some way to the story of this book. Many are maps showing where certain events occurred.

Putting the Pieces Together

The image, "Bottles and Glasses of Bubbly," was created for the shapes of the bottles and glasses, not for the subject of alcoholism. But the subject does have some relevance in this book.

The images at the end of the book are intended to complete the author's story with colorful, joyful, musical scenes.

Piece Number 1
Those Most Influential

First, I'm going to discuss the most influential people in my life, starting with my mother. My mother was born in Essen, Germany, and came to the United States in 1903 at age three. Her father, my grandfather, was a butcher. I never knew him, but I did know my grandmother. They settled in Mill Valley, California, across the bay from San Francisco. My mother had two sisters, my Aunts Babe and Betty; the first was a drinker and the second was not. From my mother's first marriage, she had a son, Frank. Later, she divorced her first husband and married my father, leading to the births of me and my brother Billy. My mother was an overly sensitive woman whose feelings could be easily hurt, and she would usually resort to drinking to heal those hurt feelings.

My father was born in the Upper Peninsula of Michigan to a large, Finnish farming family. His father, my grandfather, was a wife-beater. At one time, he beat his wife, my grandmother, so severely that he thought he had killed her. So, he went to the barn and shot himself.

His children had to drop out of school at very young ages to help their mother run the farm. My father was limited to a third-grade education. When old enough, he joined the Army and served in the Philippines. He then served in the Presidio of San Francisco— the city where he met my mother. As a civilian, he was a construction worker, a laborer. I suspect his lack of education held him back. He was always a low man on the totem pole, and that frustrated him. So, he drank when the opportunity presented itself, and that was usually when my mother started drinking. But sometimes he would stop at a bar on the way home and get drunk before coming home. He had to take a bus or streetcar to get home. I feel embarrassed just thinking about it. And his drinking would get my mother started.

My half-brother Frank was born in San Francisco to my mother and her first husband. He fought in World War Two, earned a Silver Star at the Battle of the Bulge (which I am very proud of). He married a Georgian girl. After being discharged from the army, he and his wife, Bernice, came to live in San Francisco. The first night with us, he showed us his war souvenirs: a Luger, stripes from a German soldier, Epaulets from a German Officer, a German flag, and more. real things from the real war.

He had been in the field artillery. While in Germany, he wrote my mother, "We just blew the hell out of your hometown of Essen."

Frank took a job as a carpenter and liked his work. Unfortunately, Bernice and my mother didn't get along very well, and they decided to return to Georgia. Frank lived in Georgia until his death. He was about fifteen years older than me and went into the Army when I was quite young, so we were never very close in the early years. In the later years, we got on better terms and got to know each other better. He and his family would return to San Francisco one last time. The story of that visit will follow in Piece Number 7.

My younger brother Bill, known as Billy, was born in San Francisco about five years after me. He was a little guy and liked by everyone. At an early age, he would go around to nearby construction sites, pick up the nails and other debris. The construction guys would give him a quarter, and he'd come home boasting about the job he was getting paid for. After high school, he joined the Army and served in Germany and Vietnam. In Vietnam, he was blown out of his bunk once. Mortar rounds were a big hazard. And he talked about the children, who came up to the fence to throw grenades over the fence. Such a terrible war, but they all are.

As a civilian, he worked at the Presidio of San Francisco. He married Linda and they lived in Pacifica, California. They had a daughter, Karen. He was followed in death by Linda My best friend was another Bill, and he, too, was born in San Francisco two years before me. He was my baseball pal for many years, usually catching my pitches. We both loved baseball. We were the best of friends and would usually get in trouble together. He was also my source of cigarettes until I turned eighteen, when my mother allowed it. He later got married, limiting our friendship. Then came two children, whom I babysat frequently. I was Uncle Buddy. While I lived in San Francisco, I was known by my nickname, Bud or Buddy. Then Bill got divorced and later moved to Hawaii. Bill's father was Hawaiian, but at that time, Bill didn't want me to tell anyone. Hawaiians have dark skin, and there was prejudice against them here in the States.

Bill's mother, Vi, was my second mother. We had fun together, usually playing cards along with Bill and his wife. She went to Hawaii with Bill and passed away there.

Many years later, my family and I went to Hawaii. I didn't expect to find him, but I looked in the phone book and there he was. So, we met him and his wife for dinner and had a good visit. A few years later, Glenda and I returned to Hawaii, but Bill couldn't be found at that time. Never saw him again. I wonder if he or any of my old friends are still living, but I doubt that they are. Maybe we will all meet in heaven.

I was born in 1937 and lived in San Francisco until 1958, when I joined the Air Force. I graduated from Polytechnic High School in 1955. After leaving California, I lived in Texas, Colorado, New Mexico, Oklahoma,

Texas again, and Oklahoma again. I am an Air Force veteran, a University of New Mexico graduate, and a retired computer programmer.

Last, but not least, is the most influential person in my adult life, the wonderful woman I married in 1959, Glenda. She was born in Creston, Iowa, and grew up in Lenox, Iowa, a small farming community not far from St. Joseph, Missouri.

Glenda left Lenox after high school and attended a secretarial school in Omaha, Nebraska, then took a job with the Navy Department in Washington, DC. After about a year, she and her girlfriend moved to Denver, planning to make their way to California. In Denver, where I was serving in the Air Force, we met on a blind date, and as the old song of the same name goes, "zing went the strings of my heart," and we were married a few months later.

Glenda is a small-town girl, and I am a big city boy. But that hasn't kept us from having a wonderful marriage. One time, we went to Lenox for a family gathering. To my surprise, the women gathered in one group and the men in another, including me. The topic of discussion was fuel oil, fertilizer, and the cost of seed. I sat there with a dumb look on my face. Finally, someone turned to me and asked what I did. I replied that I was a computer programmer. After a considerable silence, the conversation returned to farming. I didn't quite fit in, but the more trips I made to Lenox, the more I learned about raising corn and beans. We owned a farm in Iowa for several years, but we didn't live there. Glenda's brother grew corn and beans there and made us some money.

We have two children, Eric and Diane, and two grandchildren, Holly and Angus. At the time of this writing, Glenda and I have been together for more than sixty-six years.

Several years ago, I wrote a small book about my childhood, "Memories of a Broken Childhood and Beyond," and later, one about one of my Air Force adventures, "My African Adventure." At this late age, I looked over the years and the many events of my life and decided on "Putting the Pieces Together". And many of these pieces come from those two books.

As I mentioned earlier, my life until the age of 19 was continually disrupted by my parents' alcoholism. Although there were some good times, my childhood was a chain of painful events. The alcoholism affecting me didn't stop until my mother died in 1970.

During my fifties and sixties, my health was good, but the seventies and eighties have been downhill. When your health goes, your life changes. I made a big mistake that changed my life in a big, bad way. As I write this, I sit in a wheelchair because my legs are too weak to carry me very far, and it didn't need to happen. Hopefully, discussing this last item later will help someone avoid the mistakes I made regarding my health. For the last ten years, I have suffered because of those mistakes.

Piece Number 2
In The Beginning

We lived in the northwest part of San Francisco, not far from the Palace of Fine Arts. I remember going there in a stroller and feeding stale bread to the ducks in the pond. We were also not far from the Presidio and the Golden Gate Bridge. We could walk down to the edge of the Bay and be across from Alcatraz. The Golden Gate Bridge was to the left and the Bay Bridge to the right. I had watched the sun set beyond the Golden Gate Bridge many times.

Although this book is intended to tell the story of how my childhood was affected by alcoholism and how I survived it, I will also tell stories of interesting events not related to alcoholism, things that I remember from long ago, usually things that affected me in some way. And I remember World War Two. It was part of my early life.

I can remember two particularly memorable events that we saw from the edge of the Bay.

First was an escape from Alcatraz. We hurried down to the Bay side and saw helicopters with spotlights flying around. Several boats with spotlights searched. Prison officials said the undercurrents prevented anyone from escaping Alcatraz, but bodies were never found.

The second event was the return of Admiral Halsey's Pacific Fleet from fighting the Japanese. They came in under the Golden Gate Bridge, many, many ships. Tugboats sprayed towers of water in their honor. We could see damage to many of the ships. Carriers had blackened and twisted flight decks, and others displayed obvious fire damage. They headed to the

Hunters Point shipyards for repairs, and sailors filled the streets of downtown San Francisco.

Allow me to tell you a few more war stories.

I recall seeing many different military uniforms on the streets, especially on Market Street. And there were some scary-looking Turks with their bushy beards.

In San Francisco, we had constant air raid tests. At night, the sirens would sound. Our lights had to go off, and the black shades must be pulled down. We spent some time sitting in the dark. Outside, searchlights panned the sky. Was this the real one? I believe the Japanese had a plan to attack the city, but never carried it out.

Then came the VE and VJ days. On both occasions, there was dancing in the streets of San Francisco, with horns and sirens sounding everywhere. What a celebration!

One more war memory—this is a sad one. There was a clothes cleaner on our block, and he would put pictures from the war in his front window. There was one picture I'll never forget. It showed an American Airman in a leather jacket about to be executed by the Japanese. He was on his knees, arms tied behind his back, and blindfolded. A hole was visible in front of him. A big Jap soldier was holding an enormous sword over his head, and he had a big smile on his face, as did most of the Jap soldiers surrounding the scene. There was a tank in the background flying the Jap flag, with many soldiers standing on it, cheering. I wondered what the poor guy was thinking, knowing he was about to die. I've carried this picture in my mind since first seeing it, and it's still quite vivid.

Before we address the drinking problem, let me recount a hilarious event involving some sailors. One of our side streets was a steep hill that was great for riding skateboards down. A skateboard was a single skate with a small board across it. You sat on it and steered by tilting the board left or right. We could go really fast down our hill. One day, while we were skateboarding, three sailors in uniform came up the street. It was obvious they had been drinking a bit. They saw us riding our skateboards and came over and told us they wanted to give it a try. Then they took us to the corner candy store and bought us whatever we wanted. So, we showed them how to ride and let them have at it. What followed was so funny. They started down the hill and crashed into parked cars and the sides of buildings. When they did get to the bottom of the hill, they didn't know how to stop and crashed again. Despite the damage to their uniforms, they kept at it and had a great time. Everyone was enjoying themselves, including those guys. After another trip to the candy store, they had to leave. But it was a memorable event indeed.

As I said, I was a child of the war years. Almost everyone tried to contribute to the war effort. We, kids, collected newspapers, metal, and anything being discarded. Trucks would come by and pick up anything left on the corner. My mother would take cans of grease or fat to the local butcher. And I remember the ration books with little colored stamps. The "Greatest Generation", YES. Could it happen again, the population making huge sacrifices. Men and women endangering and even giving their lives for their country. I very much doubt it. We have seen the best of the best, and they are gone. Pardon my pessimism, but I have strong feelings about certain things that aren't part of this book.

Piece Number 3
Parental Alcoholism

Now, to the effects of alcoholism on my life. My earliest memory of my mother's drinking occurred when I was four or five years old. And there is a story that goes with it.

One day, my mother and I were going shopping on Chestnut Street, not far from where we lived. We had to cross Lombard Street to get there. At that time, there was construction on Lombard Street, and at the intersections, there were large detour signs in the outer lanes.

Just as we started to cross the street, a gust of wind blew over the sign in our lane. It landed on my head, and I was knocked unconscious. I remember someone in a uniform picking me up and carrying me to an Army car. I remember the star on the side of the vehicle. He took me to the hospital, where they took X-rays and kept me for observation. During that time, I remember the decorations for a 4th of July party.

My parents agreed to a settlement that was to provide me with $2500 at age twenty-one. I was entitled to that money, but would never get a penny of it. I'll explain later.

When I came home from the hospital, my mother put me in the big bed, my parents' bed. Sometime later, I called for her, and when she came into the room, I could tell she had been drinking by the way she looked and talked. I'm not sure how a four- or five-year-old could sense that, but I knew. There must have been earlier events leading up to me knowing when she was drinking.

When I got older, I often wondered, and still do, how a mother could bring her child home from the hospital and start drinking? What made her do that?

That was but the early years of drinking by both parents. And my brother and I suffered the consequences.

My mother was the "Doer". When they needed more wine, she got it. When we were asked to leave a residence, she would find a new one. When they needed money, she knew how to get it. When it came to wine, she had a system. Back then, you could open a charge account at any store, and there were no background checks of any kind. She would run up a tab at one store, then move to another when that tab was due. When all the stores in our neighborhood wanted their money, we'd move to a different neighborhood. When the money ran out, she would milk the inheritance left to me by my grandmother or, unknown to me, the money from my accident. Both had a stipulation that the money could be used for my well-being. And they always needed money for my well-being. The accident money was used up by the time I reached age 21. Eventually, I did get some of the inheritance from my grandmother.

I did what I could to stop their drinking. If I found a bottle, I would dump it down the sink. When they couldn't find their bottle, they accused me of dumping them, but I'd deny it.

So, what was I doing while they lay passed out in bed? Well, I had to take care of myself. And I was completely unsupervised, so I'd go through my mother's purse or my father's pants and take whatever money I could find.

I would eat at restaurants, go to movies, buy myself toys, and sometimes I'd treat some of my schoolmates to hot dogs and shakes after school. I was free to do whatever I wished. As I look back on this time, I think about how fortunate I was not to have been molested by anyone. I went to the movies late at night and ran around in the dark city streets. I wonder what would have happened to me if I had been a girl?

Another story to tell. I think it was on my eighth birthday that I received a brand-new Schwinn bicycle (with my parents' approval). What a beauty! My father taught me how to ride it. So, one day I took it to school, put it in the bike rack, and went to class. When I returned, it was gone. I went to the nearby police station and reported the theft. An officer took me to a big garage in the back and showed me in. It was filled with hundreds of bicycles. He told me to look around for my bike, but I couldn't find it. He told me to keep coming back, and maybe it would turn up. Week after week, I checked out that place, but to no avail. Finally, the cop said to me, "Take one, any one you want." I think he felt sorry for me. So, I picked out a good-looking one and rode it home. Thanks, Mister Cop! It feels good to talk about the good people I encountered along my way.

When my parents were drunk, they were mostly passed out in bed. But mostly means not all the time. And that was when there was trouble. My mother was usually a sloppy drunk, but she could be very argumentative if she wanted. But when only slightly drunk, she could be very mean and nasty, especially to my father.

My father was a very quiet man when sober, but also became mean when he was drunk. As a result, there was a lot of fighting. There was a lot of vulgar name-calling, things a young child shouldn't have to hear. And there was physical punishment for my mother when she said the wrong thing, as she frequently did. My father, a construction worker, was not hesitant to hit my mother and often did so when drunk. I've seen her broken glasses on the floor many times—and even worse, I've seen many black eyes. On several occasions, her screams would cause neighbors to call the police. They would come and calm things down. This was very embarrassing for me. The neighbors always came out to see what was going on: those damned Jacobsons again.

One evening, I was out on the street playing with a bunch of neighborhood kids in front of our building when a police car with a siren blaring came racing down the street. It went past our place, then stopped and backed up. I knew where they were going, to my flat. I was so ashamed that I ran into the backyard and hid under the steps until the cops left and the neighbors went back. No one was arrested. But having this happen in front of all your friends was so embarrassing.

One time after some fighting, my father was afraid the police were coming, so we left our flat and walked a few blocks away to a vacant lot with an advertising sign on it. There was a platform behind it where the streetcar conductors would eat their lunch. So, my father and I slept there with only newspapers for covers. It gets very cold in San Francisco at night, and sleeping under newspapers is not a good thing to do.

On another occasion, the police came into our flat and looked around. My father kept telling me to hide under the bed, but I was too scared to do it. This led to the first time my brother and I were taken away from our parents and placed in the Juvenile Detention Center.

It was not a nice place for innocent five- or six-year-olds. I was quite young and remember seeing my baby brother in a bassinet behind a big window. I must have been around six. Other than that, I don't remember much about this first visit, except that a social worker would come and play games with us. I was told that my parents would have to go to court to get us back. They did, and we returned to them.

One would think that they got the message, but they didn't. And our return to jail occurred a few years later, and it was much more memorable. The judge was not as lenient as the first time.

So that was my life until about age nine, when things really took a turn for the worse. Another series of drinking events led to our second visit to the San Francisco Juvenile Detention Center, the place where there were bars on the windows and matrons to watch over you. Where you lived in a cell-like room. Where the door is locked every night, and a matron comes around every few hours and shines a flashlight on you to make sure you are in bed. Where the rooms are lettered, and the 'H' room, or 'Hell' room, was a padded cell for those who misbehaved.

We, the twelve and younger, were located on the eighth floor of a ten-story building in downtown San Francisco. The two floors below us were for teenage girls, and the two below that for teenage boys. While I was there, some teenage girls overpowered a matron and escaped. So here on Otis Street in downtown San Francisco began a multi-year journey with many stories to tell, some not so good and some very good.

I'll start with how the second return to the Detention Center occurred. My mother's sister had a nice home in Oakland, and for some reason, it was agreed that we would move in with her and my two cousins. My aunt was also a heavy drinker. One Saturday, Billy and I returned from a movie to find all hell breaking loose. My aunt wanted us out of her house and was very vocal about it. I never knew why. She said something about the police, so I got scared and went all the way back to San Francisco.

I went to my best friend's flat and asked his parents if I could stay with them overnight, and they agreed. But they knew I couldn't stay with them. The next morning, a lady from the Juvenile Authorities showed up and took me into custody. My friend's parents said there was nothing else they could do. I understood and went off with the lady. The authorities also got hold of Billy and took him into custody. We would not see our parents again for almost two years. And many things happened in those two years.

It was here that I first had feelings for a girl, but a much older girl. The teenage girls below were allowed to come up and keep us, the twelve-and-under kids, company. They played games with us, read stories to us, and we did puzzles. This one girl took a liking to me and would always come to me first. After a while, she told me that she was going to take me home with her when she left. And I believed her. And I did the second time she told me the same thing.

The day came when the girls came up except for her. She had gone home. No goodbye, nothing. I never saw her again. She broke my nine-year-old heart.

Piece Number 4
Polio

One morning, I woke up with a bad headache and a sore throat. Then came a fever. I was put in bed and later had a doctor check me out. The following morning, three doctors showed up to check me out. I was told I had polio.

Later that day, two ambulance attendants showed up with a stretcher consisting of two poles and some canvas in between. They got me to the first floor. As they carried me to the ambulance, I remember the feeling of freedom. Being out of that place felt good despite my illness. And I got a ride in an ambulance to the hospital, San Francisco County Hospital.

More stories to tell. At the hospital, I was placed in an observation room which had a big window to the hallway and another observation room, also with a big window, across the hall from me. As I was lying there, they brought in another small boy into the room across the hall from me. He was accompanied by a doctor, a nurse, and apparently, his parents. It looked like they were doing something to him, and he was having a fit, screaming and yelling. Other nurses came in to hold him down. They had a hard time, and he was still screaming.

After a bit, a nurse came into my room, and I asked her what they were doing to that boy. She told me they were doing a spinal tap. So, I thought a spinal tap must be some form of painful torture and was thankful I'd never needed one. They eventually got it done over there. And then some doctors came into my room. They examined me thoroughly and talked among themselves. And then I heard the words "spinal tap" again. I was panic-stricken. They were going to torture me. Then they left. I hoped my ears had deceived me. But shortly, a man came into the room with a tray full of instruments. I was really scared now.

He told me he was going to do a spinal tap. I may have wet my pants at that point. He had me lie on my stomach. Then he told me I'd feel a small prick in my lower back. And so, the torture would begin. I felt the small prick, and after a bit, he started doing something back there. A few minutes went by, and then he said, "All done." What? Where was the torture? The screaming? He took his tray and left. Wow! I guess that boy was afraid of needles. And so, my stay at San Francisco County Hospital began.

The spinal tap must have confirmed my polio because I was placed in a ward with five other boys with polio.

The treatment I received for my polio was quite simple: hot packs, twice a day. Hot packs are steam-heated chammies placed on the neck, shoulders, chest, back, upper arms, and both thighs. A piece of Naugahyde was pinned over the chammies to retain the heat. Then I had to lie in bed for an hour.

The test for being cured was for the doctor to hold my legs down at the knees and for me to sit up and touch my toes. The doctor said that when I could do that, I could go home. At first, I couldn't sit up at all. But slowly, I made progress toward touching my toes.

Let me tell you about some very kind and generous people, the parents and friends of the other five boys in my ward. At visiting times, the families and friends would bring candy, comics, books, games, and toys for their boy. Each bed would have a crowd around it, except mine. No one came to see me. Well, one time, a social worker came by and asked how I was doing and left. Someone must have asked a nurse about the kid that no one visited. Soon, five families were bringing me all those things, and they would come over and talk to me. I am always touched when I think about those kind people. There were a lot of them along my way to the future. Bless them.

The best part about being in the hospital was Miss Ely. This hospital trained student nurses, and one of the new ones that started while I was there was Miss Ely. She was the most beautiful nurse in the world. She was twenty-one years old, quite petite, and had the most beautiful face. In those days, nurses wore all white: a white cap on their heads, a white dress, white stockings, and shoes. She was so beautiful. And she took a liking to me, probably because of my situation. When she came on duty, she would always come in and see and talk with me, and she did the same before she left in the evening. And the other boys were all jealous of me. One day, she asked me to be her boyfriend. I didn't say no. Now, when she came in or left, she would kiss my forehead or cheek. Wow! Did the love light shine! Now, at ten, I really had feelings for her.

But the day came when I could touch my toes, and they would let me leave the hospital and return to the Detention Center. I told Miss Ely I was very sad to leave her. She replied that she would come and see me when I returned for therapy. Another broken heart. I never saw her again.

Then the next big step in my life occurred. And it was a good, happy step. My brother and I were moved to a place called 'Edgewood Orphanage', still in San Francisco. Not really an orphanage anymore, it was home to a lot of children from broken homes. It was located in a nice residential area of San Francisco. And my mother's second sister lived not far away.

There were six so-called cottages. We, Billy and I, lived in number six, 'Williams' cottage. The cottages were modern two-story buildings. My cottage housed up to ten boys on one side and up to ten girls on the other side of the upper floor. The house mother's apartment was between the two sides. The downstairs included a living room, library, dining room, a kitchen, where I frequently washed or dried the dishes, a locker room for boys and one for girls, and an activity room. And there was a basement with additional rooms.

We had a regular house mother during the weekdays and a different house mother on weekends. Our cottage housed children under thirteen, except for one fifteen-year-old girl, Betty.

There were two bedrooms on each side of the upstairs floor, five to a room. And there was a large shower and toilet area.

Apparently, the kids knew something about me before I got there. One kid asked me if I was the new sick kid.

Because of my polio, I was given special treatment. At the public school we all went to, just two blocks away, I was allowed to leave an hour early, so that I could go to the Edgewood infirmary. There, I got a hot bath, and an hour or so in bed, plus milk and cookies. Not a bad deal!

Edgewood was probably the best place I had ever lived to that point. It was a modern, clean home, and I liked it. And the house mother was very strict with us. She gave us chores to do every day. At first, I didn't like all the strangers, but we all became friends quickly.

So many good things happened at Edgewood. Each cottage had sports teams, and we competed with each other. I was the captain of the softball team. We'd play games after school. The fifteen-year-old girl was my best player. She was a very good hitter and fielder. We won often.

I played on the basketball team, but not very well. In addition to the softball field and basketball court, we had an archery area, a horseshoe area, a soccer field, and several tennis courts.

Billy and I lived there for over a year, and we experienced several special events. Let me tell you about them.

Outside organizations did much for us. They sent buses to pick us up and take us to the downtown theaters, where they had special movies for hundreds of us kids.

One time, they took us to a play. I'd never seen one and was amazed at what I saw and heard. The play was 'H.M.S. Pinafore' and it was the most colorful thing I'd ever seen. And the singing and dancing were wonderful. "I polished up the handle so carefully that now I am the ruler of the Queen's Nivy." I will never forget those words.

And I must tell you about Christmas and, once again, about the generosity of many kind people. At Christmas time, one of the civic groups, like the Moose, Elk, or Kiwanis, sent a man by to see what we wanted for Christmas. One at a time, we were allowed to tell the man what we wanted for Christmas. Before going further, let me tell you that at some previous time, someone went to my parents' place and got some of my favorite toys. One of my favorite things was a 16 mm projector from a previous Christmas. At Edgewood, the house mother would let me show movies on the wall for the other kids. I had a bunch of cartoons like 'Felix the Cat' and 'Barney Google.' I told this man that I wanted another cartoon for my projector. Just before Christmas, our boxed gifts arrived, and I was surprised by the large box for me. Why is it so big for a movie reel? When I opened my box on Christmas day, I found six cartoons in it. Wow! I never expected such generosity. They made at least one kid very happy.

Then there were the people who lived around and near Edgewood. Just before Christmas, three of us kids went to a neighbor's house for Christmas with them. They had bowls of candy everywhere. They gave us presents and a big turkey dinner. It was a very merry Christmas for three kids from broken homes. And more neighbors did the same for the other kids. I bless these kind people for their generosity.

And there was more. I was invited to attend the 'Shrine East-West' All-Star football game at Kezar Stadium. It was very exciting.

On another occasion, I was invited to a 49ers game by a man named Pat Brown, the District Attorney of San Francisco and future governor of California. I remember some of the names of the players I saw: Frankie Albert, Joe Perry, Hugh McElhenney—all Hall of Fame members now.

At Edgewood, I got to do things I never would have had a chance to do otherwise.

I did have one scary experience at Edgewood. One school morning, I had gone to the dentist first, so I didn't get to walk to school with the other kids. The streets were deserted now, and as I walked along, a car pulled up to me. A man in a suit opened the window and asked where I was going. I told him that I was going to school. He offered me a ride, and I declined, saying, No, thanks. Then he got insistent and kept telling me to come closer and talk to him. I slowly walked closer to the car. When I got closer, he opened the door, and that scared me, so I backed off quickly. He growled at me, slammed the door, and left. I wonder what would have happened to me if I had accepted that ride.

One last unfortunate incident at Edgewood involved my brother Billy. When we got to Edgewood, Billy was having a problem with bed-wetting. The regular house mother tolerated it, but not the weekend house mother. The bed-wetting happened several times under her watch, and she didn't like it. She was a tough woman, and she'd had enough. And down the stairs he came, forced to wear a diaper. He was humiliated, and I felt so sorry for him. I don't think his bed-wetting was intentional, but rather something he couldn't control. I don't think this was or is an acceptable way to solve the problem. Because of this treatment, Billy never had anything good to say about Edgewood.

And so, it came to pass that my brother and I were reunited with our parents after almost two years. We were allowed to go home with them on the weekends. They had a small apartment, but there was enough sleeping room for all of us. They weren't drinking, and they seemed to want us back. I guess they had been sober for a long time, and that's what let the authorities reunite us. So, the weeks went by, and we spent the weekends together.

To be honest, it was nice to be with them again. Things were working out well for all of us. Eventually, we were allowed to leave Edgewood and return to our parents permanently.

What happened the day we left Edgewood is unbelievable. It just couldn't happen.

My brother and I gathered up all our things and were waiting in the living room for our mother to pick us up and take us home. The house mother waited with us. Eventually, a taxi pulled up in front of the cottage, and my mother got out. Just one look at her face almost brought me to my knees. This just couldn't happen. What was the house mother going to do when she realized that my mother was DRUNK?

I can't express the terror I felt. This just couldn't happen. Her speech was slurred, and her face told everything. But our house mother never noticed and let us go with her. Back into the belly of the beast, back into Hell, more years of it. God help us.

Piece Number 5
The Junior High School

I was now thirteen and ready to enter junior high school. I attended junior high, made the baseball team, and had several good friends at school. My three years in junior high were pretty much spoiled by long periods of drunkenness with a few occasions of sobriety. But overall, it was a bad time for Bill and me. More fighting, more evictions, more not-working, more no-money, more anguish for me.

One evening, I had been out with the guys until late. My parents had been drinking, and the apartment was a pigsty, food out everywhere. On walking in, I noticed a business card on the table, and I was sorry I looked at it. Someone from the Juvenile Authorities came by to check our situation. If he saw this mess and my parents' condition, my brother and I would be doomed. I wanted to stay with my friends, to play ball for school, and to lead a normal life. It seemed certain to me that we'd go back into Juvenile custody and probably never see our parents again. Even a thirteen-year-old boy will cry, and I did. I prayed all night that they wouldn't come and get us.

And you know what? Nothing ever happened. That guy should have reported us, but he didn't. Maybe it would have been better to go back to the Juvenile Authority because the remainder of my junior high years were bad. The drunkenness continued. Baseball became the love of my life, and at this age, I got to play a lot. Fortunately, all my friends loved it, too. So, despite what was going on at home, I was playing baseball.

My junior high baseball team, the Everett Owls, won the city championship, and we had an awards assembly at school. Our coach was able to get a guy named Bobby Brown to present our medals to us. At that time, Bobby Brown was the starting third baseman for the New York Yankees. It was a thrill to shake the hand of a real ball player and a Yankee at that.

Time for another story. On my way to school, I would stop at my best friend Bill's house. He went to a different school and didn't care for school too much. So, he would try to talk me into playing hooky, and he succeeded many times. He would call my school and pretend to be my mother and give an excuse for my absence. And I would do the same for him. It worked many times, until I got caught.

We would play Canasta for hours, go shoot hoops at the park, and sometimes go to a movie. But we had to be careful as truant officers would be roaming around. Some theaters wouldn't let us in because, at our age, we should be in school. But we did it time after time. And I got As and Bs while Bill got Ds and Fs. But it caught up with me. Somebody from the school went to my house one day and talked to my mother, who was sober at the time. I caught a bit of hell over that. My mother told my counselor that Bill was a bad influence on me.

Bill was my best friend, and we were together all the time. When I pitched, he was my catcher. Pitching was my favorite position. Bill and I were inseparable. Until he met Georgia, whom he married and had two children. I became a babysitter, known as "Uncle Buddy". While living in San Francisco, I was known by my nickname, "Bud" or "Buddy". But the three of us still got together and did things.

One summer, the three of us went up to the Russian River to play in the water. None of us knew how to swim. I was out in the water at this popular beach, walking in water about chest high. Suddenly, I stepped into a hole, a deep hole, and went under. I struggled to get to the air but couldn't—the end of me. Suddenly, an arm wrapped around my throat and pulled me up into the air. The lifeguard saw me struggling and saved my life. That was too close.

When I got to high school, there was a policy that all boys had to learn to swim. During WW2, too many sailors and other military personnel drowned because they couldn't swim. So, twice a week, before school started, I had to go to a pool not far from school. And the pool wasn't heated. You can't imagine how cold that water was. We had no choice but to jump into the ice-cold water. I kind of learned to swim, but was never very good at it.

Piece Number 6
More of the Same, Then High School

My family situation worsened as the drinking continued, and we moved from place to place. There were occasions when I returned home from school to find bill collectors waiting for me, a young teenager. They would jump out of their cars and threaten me with going to jail if the bills weren't paid. I learned not to go home right after school.

I managed to graduate from junior high school and started high school. Unbelievable as it may sound, my parents' drinking stopped. And it stopped all through my three years in high school and a year working, and part of my first year in college. Can you believe it? Over four years of normality, security, happiness, sobriety. Truly amazing. No more drinking? Forever? No, this was not the end of the drinking. The worst, the very worst, was yet to come. And it will pain me greatly to write about it.

But first, let's hit the highlights before the lowlights. I entered Polytechnic High School, got good grades, played baseball, almost learned to swim, and had good friends.

Sports were part of my life, playing baseball in junior high and high school. Being left-handed limited the positions I could play, but I got to do what I liked best, and that was pitching. I loved baseball, but wasn't very good at it. As an adult, I played a lot of softball in different leagues. I also did bowling and golf. Okay in one, not so good in the other. I'll let you guess which was which. The last few years, it's been Wii bowling and golf.

Having a son and a daughter allowed me to get into coaching. Baseball with my son, softball with my daughter. I really enjoyed coaching the young kids, teaching them something I loved to do. One of my girl' teams won the City Championship in their class. That made me, my coaches, the girls' parents, and the girls very happy.

So, back to my dismal love life. I had a big crush on a friend's sister. She had a very attractive face and figure. I had a couple of dates with her, but it never worked. And there was another girl after her, who dumped me after a few dates. Strikeout after strikeout. That's me. I just wanted someone to care about and someone to care about me. But it wasn't going to happen until much later.

Another story: One evening, my friend Bill and I were playing catch on the sidewalk in front of his place. Ronnie, a friend of ours, drove up in a car and invited us to go see some girls. So, we jumped in and we were on our way. Along the way, we passed a police car, and the officer inside gave us a strange look. And Ronnie took off. "What are you doing?" We screamed and begged him to stop. But he didn't, and the cop chased us. We were driving through the narrow streets of San Francisco with a hot police car chasing us. We finally got stopped and arrested. It seemed our friend Ronnie had his uncle's car without permission. We got to spend the night in the Juvenile Detention Center, a newly built one.

Of course, my mother was notified at 11 pm and had to come to the Center. The first time she saw me, she slapped me across the face, hard. I think she was angry.

We spent the night in jail. But Bill and I were innocent. And the next morning, they recognized our innocence and released us and told us there would be no arrest record. Ronnie got to stay a little longer.

Much later, my friend Bill and I got interested in Astronomy. We attended lectures, frequented the Planetarium, visited Lick Observatory south of San Francisco, built a six-inch reflecting telescope, and joined the

San Francisco Amateur Astronomers. We really got into it. In high school, I wanted to be an astronomer. Starting college I wanted to be a physicist, but I eventually got a degree in math. I would never work in either of these fields. I quit playing baseball in my senior year to concentrate on my grades to get into college. I graduated in 1955 and started college at San Francisco City College. I overloaded myself with Chemistry, Calculus, German, English, and Astronomy. It was too much, and my grades weren't very good for two semesters. I quit school for a year and went to work. I got a job as a cost accountant for an electric transformer company. I calculated the cost of an assembly based on the costs of the parts. I saved up my money and bought a car, a 52 Plymouth. And I finally had a real girlfriend who loved me. I went back to school and was getting good grades. Everything was going great. I was very happy with my life.

Piece Number 7
The Visit

College final exams were coming up. My older brother Frank and his family were coming from Georgia for a visit, and everything was looking good. He, my mother's firstborn, was always very special to her. She was excited, and it was a happy time for all of us.

My brother and his family arrived. A couple of days into the visit, I had classes most of the day and was gone during the events at home, so I don't know exactly what happened.

It seemed my brother and his family were going to visit my aunt, my mother's sister, the one who lived in San Francisco. For some reason, they expected my mother not to go with them, but to stay and cook lunch for them. So, they left for the visit. This apparently hurt my mother's feelings. And her recourse was, of course, the bottle. And that was exactly what happened. She went out, got a bottle, and by the time my brother got back, she was passed out in bed. My brother, who had been through the drunkenness experience years before, was furious. He and his family packed up and left that same day. What he left behind was a disaster in the making.

And what followed was the worst of the worst. My father came home and said nothing. The night and the next day passed. I had classes, so I wasn't home. My mother renewed her drinking while I was gone, and, of course, my father joined her. No work, no more paychecks, no more money. Only Hell was to follow.

Another downward spiral began, and this was probably the worst one.

Studying for finals was impossible. I took them without preparation. The final grades in all my classes were poor. I cannot tell you the despair I felt. If there had been a gun in the house, I might have used it on myself. In a moment of stupidity, I thought my being unconscious might straighten them up. So, I swallowed a bottle of aspirin. My stomachache lasted for days.

The rent didn't get paid, and the landlord, who lived right above us, did not like drunks who didn't pay the rent. He told me we had to get out. I stalled as long as I could.

My best friend Bill knew what was happening and kept telling me to get away, to join a service, and get away before I go crazy. And I did just that by enlisting in the Air Force.

When the time came to leave, I parked my car in the street, left a note, and was gone. But in doing so, I had abandoned my younger brother, a guilt I have never overcome.

One would think that being in the service would protect me from what was going on in San Francisco. One would certainly be wrong.

Piece Number 8
In The Air Force

I was in the Air Force now. After some basic training at Lackland AFB in San Antonio, I was sent to an Air Force school at Lowry AFB in Denver. I was studying electronics, a new and difficult subject. One day, I was sitting in class, and a sergeant came in wanting Airman Jacobson, me. He told me that I had a phone call. "Who's calling me?" Well, it was my best friend, Bill. He told me that things are bad with my mother, father, and brother. They were evicted, and he was not sure where they were living. Father had fallen and gone to the ER with my brother. Mother was in the psycho ward in the hospital. And someone needed to fix things. That someone was, of course, me.

They were disrupting my Air Force career. I told Bill to contact the Red Cross, which he did. My squadron commander called me in. He was not happy with me leaving my class at this time. But he let me go. The Red Cross loaned me money, and I was on a bus to San Francisco.

I got there and stayed with my friend. I found out that my father's family in Michigan had sent him money to fly him and my brother to Michigan. And that was where they were now. I didn't know how that came about. That problem was taken care of, now to my mother.

Somehow, I found out where she was and what was going to happen. She was in the psycho ward at General Hospital. The next day, a judge would be there to consider her situation. I had to be there, and I was.

The psycho ward was a huge room filled with beds, my mother's included. I found her, and she begged me to get her out of there. She told me there were crazy people there.

Shortly after, the judge appeared, looked at some papers, then turned to me and asked what I wanted to do with her. I hadn't expected that. I asked what options there were, and he kind of grunted and told me she could be released or sentenced to 60 days here. She pleaded with me. I couldn't let her out; she had no place to go, no one to take care of her. So, I said to keep her here, and the look on her face changed to hate. What could I do? I couldn't stay here any longer. I returned to Denver, but had I done the right thing?

I got back to Denver, and they let me catch up on the missed classwork, which I did. I got to stay in the same class. The Air Force was first teaching me electronics and then how to maintain a device used in atomic weapons. It took me a year to get it done, but I did get it done.

And do you know how many times I worked on that system after that? You're right – zero! After graduation, I never saw that device again. But electronics would prove useful in what was to be a very good job.

And during this time, it happened. I met a girl, Glenda, in Denver, on a blind date. The date was a picnic in the Rocky Mountains. After one date, I decided she was special. I guess she liked me, too. We were married a couple of months later. As of this writing, we have been together for over sixty-six years.

Shortly after our marriage, Glenda continued a correspondence course in oil painting she had started before dating me. It looked like something I wanted to try. And I did, and I liked it. Since then, I have been tied to art, especially digital art on my computer.

Shortly after our marriage, I graduated from the Air Force school I had been attending and was waiting for my next assignment. When they were posted, I was scheduled to go to Taiwan. This was terrible news for us. I could be there for a year or more, and Glenda couldn't go with me. It was a very sad day for us. The next morning, I went to class to find that new assignments had been posted. And I had a new assignment, the Air Force Special Weapons Center in Albuquerque, New Mexico. So, we packed up the car and a trailer and headed south.

After searching for an apartment for some time, we found a nice apartment not far from the University of New Mexico. It had a living room, bedroom, kitchen, and bathroom. And when the lights went out, cockroaches. The owner was an elderly woman who treated us well. We would live there for almost six years.

We found the "Land of Enchantment" to be just that, enchanting and quite different from other places either of us had lived. There were many sites to be seen in New Mexico. We visited several Indian pueblos where we saw many native American rituals, dances, traditions, and habitats. We visited Bandelier, an ancient, long-abandoned pueblo site.. Bandelier was a perfect place to take visitors, and we did take many there. We made our way to the top of the "Acoma Sky Village". We saw the Petroglyphs Monument and the Gila Cliff Dwellings. Of course, we made many trips to Santa Fe, Taos, and Los Alamos. And there was much more in Albuquerque, Old Town, the "National Museum of Nuclear Science and History', where you can take a look at "Big Boy" and "Fat Man", and Kirtland Air Force Base. Then there was the Sandia Tramway up the side of the Sandia Mountains. Finally, there was the food of New Mexico, which was wonderful. I almost forgot the "Hot Air Balloon Festival". In the southeast corner of New Mexico is Carlsbad Caverns, a site not to be missed. There is also great trout fishing in the mountain streams in the north. When in Santa Fe at over 5,000 feet, you feel like you can reach up and touch the clouds. And there were the native American gifts and jewelry stops along the roads and highways. The beautiful, handmade silver and turquoise jewelry was worth the stop. The "Land of Enchantment", yes indeed. And don't forget the beautiful artworks in Taos.

Well, back to my parents. Things eventually changed for the better. My mother and father were reunited and lived in a cheap hotel, which my mother managed, so my father didn't have to work. His heart was a problem. But they lived and managed there for several years. There was no drinking, and my brother Bill was nearby in Pacifica.

My first assignment at Kirtland was mostly doing nothing. We sat in a hangar reading magazines and did an occasional cleanup detail. Then I was assigned to a "Sampling Squadron". The squadron had several jet aircraft that could sample the air by flying through or around a mushroom cloud. The aircraft had special paper filters in the wingtip pods. They also had tubes that sucked air into a steel bottle in each wing.

Since we were not testing in the atmosphere at the time, not much was going on. I was assigned to test and maintain Geiger counters, of which we had dozens. I did get in on one mission at Indian Springs AFB in Nevada. They were testing a device that omitted radioactive exhaust.

During the test, our plane sampled the air around the exhaust of the device. Our job was to recover the filters in the wing pods and the steel bottles in the wings. We also had to recover the pilot with a forklift. Didn't want him to touch the outside of the aircraft. I had a very important job. I stood downwind of the aircraft with a pole with a hook on the end. If one of the paper filters got blown by the wind in my direction, I would hook it. On the flight back to Kirtland on a C-47, I got airsick for the first time. I was glad to get off that airplane.

Putting the Pieces Together

After a few months in that squadron, I was reassigned again, this time to a telemetry shop. It provided me with the best job I had in the Air Force. The shop had both military and civilian personnel. And they had plenty of work to do. So, I became a telemetry technician. My knowledge of electronics got me a good job. In this shop, I learned to assemble, test, and operate telemetry equipment for projects involving atomic and nuclear weapons and their possible deployment. After some time, I became quite experienced in telemetry and was selected to work on a special project.

Piece Number 9
My African Adventure

The project involved sending a telemetry package over the South Pole to investigate the number of alpha particles, beta particles, and gamma rays accumulating at the pole. Large quantities could affect the electronics in anything sent over the pole. In preparation for this project, special, highly sensitive receivers were purchased from a company in California. I was the first one chosen to become efficient in the operation of them. On one occasion, the captain in charge of the project and I visited the company in California.

A short time later, a space electronics show was scheduled for Albuquerque. The company from which we bought the receivers was going to have a booth there and planned to show their new receivers, but their equipment was delayed in getting to the show, and they had nothing to show on the first night. So, they asked us to provide our equipment, which we agreed to. In addition to the hardware, they wanted someone from the Air Force in the booth. Guess who? They paid me $100 an hour for four hours with them. Best job I'd ever had or would have, money-wise. But it was difficult to talk with space engineers and other experts in the field. They asked difficult questions.

The preparation for our project continued until it was time to determine where to locate the receivers. They needed to be in the southern hemisphere, so Australia, Hawaii, and South Africa were chosen, with South Africa being the choice assignment, my assignment.

I had not been out of the States except for a brief trip to Juarez, Mexico, and now I was going all the way to Africa. Wow! Being an Airman First

Class gave me little authority to do anything, so a first lieutenant, an electronics engineer, was chosen to go with me. So, my Lieutenant, Kenneth, and I, traveling in civilian clothes, set out for Pretoria, South Africa, where the United States had a satellite tracking station, part of the Atlantic Missile Range. They had a 35-foot dish antenna there.

We flew out of Albuquerque, stopped at McQuire AFB in New Jersey, and eventually arrived at Rhein-Main Air Force Base in Frankfurt, Germany.

We had an evening to spend in Frankfurt, so we went to a German Bier Garden. There was a happy show on the stage and plenty of beer going around. And, of course, I overdid it on the beer and was loopy when we left.

We took a taxi back to the base and let Kennith off at the Officers' Quarters. Then to the Airmen's Barracks for me. It was up to me to pay the taxi fee. I gave the driver a few marks, but he kept saying "More, more". I gave him all the money I had, and he left with a smile on his face.

The barracks were a huge room with many double bunks in it. When I arrived, the room was empty, but when I woke up the next morning, it was full. An excess of German beer can help you sleep undisturbed by anything.

The next day, our goal was to find a South African Airways flight to Johannesburg. This was not easy, but it did get us a small tour of Europe.

From Frankfurt, we flew to Zurich, Switzerland, where we had a day and a half layover. We went to a restaurant and found the menu to be written in French. But I happened to see the word "beef" in one of the choices, so we ordered that, and it turned out to be a good meal. A woman with a small dog came in and sat in the booth across from us. When they brought her a meal, they also brought a plate of food for the dog. Never saw that at home.

After an overnight stay, we went to the airport to wait for a flight to somewhere. While sitting there, I heard a page for "Herr Yacobson und Herr Warley". At first, I didn't think anything of it. After the third time, I told Kenneth that I thought we were being paged. And we were.

Next, we flew to Rome, where we had a stop for a plane change. I saw many priests in the terminal for some reason.

On to Athens, Greece, where we finally got a flight on South African Airways to Johannesburg. And we were on our way south.

Our flight flew over low clouds for a long time. There was nothing to see below. Then we came to a large clearing, and I could see a lot of green down below. And I realized that I was really flying over Africa, Africa with lions, elephants, and crocodiles. The same Africa that I had read about in books and seen in movies. The land of Tarzan, Jane, Boy, and Chetah. Then we landed at Salisbury, Southern Rhodesia, for a stop, and I could see natives in costume, dancing to their music, just outside the terminal. Yes, this was Africa.

Finally, we arrived in Johannesburg, where we rented a car for the drive to Pretoria. We rented a nice little red sports car, but found that the steering wheel and gear shift were on the wrong side of the car. Oh, oh! Could it be that these people drive on the wrong side of the street? The answer was "Yes". I let Kenneth do the drive to Pretoria, and he got us there safely. But my turn would come, and I would become very good at driving in the left lane. South Africa at this time was in Apartheid. Blacks were at the bottom of the social list. They lived in a ghetto and worked for low wages.

Pretoria was a beautiful city. The white residents had plenty of cheap help. Gardeners to keep the outside looking good, and maids, housekeepers, and servants to keep the inside looking the same.

We arrived in Pretoria, checked into our hotel, and decided to have a drink before dinner. We went to the bar and ordered our drinks. When we finished, we paid our bill, left a customary tip, and went to dinner.

The next day, after spending the day at the tracking station, we returned to the bar for another drink. When the black servers saw us come in, they practically ran to be one to serve us. And the first one there did. I asked a gentleman sitting not far from us, "Why the rush to serve us?" He asked if we were good tippers. "I guess, we left a 15% tip last night." "No, no," he replied. "Don't tip that much. Only pennies, just leave pennies. You'll spoil them with bigger tips." But we still tipped more than pennies.

We came to know the "boys". They were of the Bantu tribe and were very nice people. Our server at the restaurant was named Robert. He was an older gentleman. We got to know him quite well. He would joke with us and we with him. We had a good relationship with him and the other "boys" as well. That evening, I asked the desk for a wakeup call at 7 am.

At 7 am the next morning, there was a knock on my door. I invited the knocker in, and in came a little old black lady carrying a tray with a teapot and a cup on it. "It's me Masta", "It's me, Masta," she repeated and set the tray on the nightstand. I thanked her, and she left. What a great way to start the day with a cup of fresh-brewed tea. But I didn't need the "Masta" part.

There wasn't too much to do at the station once our receiver was put into the equipment rack. We had a stack of tapes ready to record the received data. We'd drive to the station daily to see if anything was happening. On one of our drives out there, we came across a dead donkey on the side of the road. It stayed there for days, and each day, the site and the stench got worse. We could smell it a mile away. You didn't want to look at the rotting carcass as you drove by. Finally, someone poured gasoline on it and burned it. The stench was gone, but the burnt carcass remained. Not a pretty sight.

On this same road was a hamburger joint shaped like a Flying Saucer. We stopped there for lunch one day and ordered hamburgers. It took a long time to get them, and when we did, I found hair in my burger. I wondered what they butchered to get that meat.

Kenneth was needed in the States to help with the rocket setup and launch. So, I drove him to Johannesburg and saw him off. By now, I was doing good at driving on the left side. But the money system was still a problem. Rand, Pounds, Shillings, Pence, and more, converting these to dollars in my head was not easy.

The missile launch was delayed several times, so I had time to get around Pretoria and see the sights. And because I was in the southern hemisphere, November was a summer month.

One of the interesting sights I visited was "Voortrekker Monument". The monument was surrounded by a circular wall on which were depictions of covered wagons. South Africa has a settlement history similar to that of the United States. While our settlers headed west and fought Indians, the English and Dutch settlers headed south and fought the Zulu. Movies have been made of the epic battles with the Zulus.

The Presidential Palace in Pretoria houses the Executive branch of government. The Legislative branch is in Cape Town, and the Judicial branch is in Johannesburg.

Once Kenneth left and I was alone, my seating in the dining room changed. I was always seated at a table with other people, usually Europeans. Some would talk with me, and some wouldn't. Very uncomfortable sometimes. It was not difficult to separate the Yanks from the Euros in the dining room. How we use our silverware gives us away. We, Yanks, used the knife in our strong hand to cut, then switched the fork to our strong hand and the knife to our weak hand. Euros don't switch them and use the knife to push food on the fork. Crazy Yanks!

One evening at dinner, there was a man alone at a table, and he was berating the server for the small portion of meat on his plate. He was loud and, obviously, from his speech, an American. Any other Americans, like me, had to be embarrassed by his behavior. But he went on and on, as if the server were responsible for what was on his plate. Finally, he got up and stormed out of the dining room. Indeed, the "Ugly American" did exist and maybe still does.

After coming to breakfast alone a couple of times, Robert asked me where my brother was. I told him that he wasn't my brother. Robert had a puzzled look on his face and said, "All boys think you brother-brother," then added, "Both have no hair on top." Both of us were thinning on top and were of about the same weight and height. We could be mistaken for brothers.

One morning, I thought I'd try something different for breakfast. I looked over the menu and saw something called "pelches," which sounded like peaches. So, I ordered it. Shortly, a plate was set before me with a big eye looking up at me. A whole fish covered in tomato sauce. I abandoned my table quickly and hurried back to my room. What was such a thing doing on the breakfast menu? Another specialty on the dinner menu was "Monkey Gland Steak," something I had never tried. One thing that I enjoyed was the tomatoes. They had a wonderful flavor, and I remarked to Kenneth that they must grow the tomatoes in elephant dung.

Finally, the day came for the launch of the four-stage missile. I waited at the receiver, hoping to acquire the signal and to record a lot of good data. I waited and waited. Then someone came into the room and shouted, "Wrap it up. It's done." We were told that the third and fourth stages failed to separate properly, and the package was tumbling too low for us to see it. "Damn!"

So, we boxed up the equipment for shipment back to the States. And I packed my suitcase and started thinking about getting home again. I had been in South Africa for a month. And I never did see a cobra at the station, although some of the "boys" did carry around poles with blades on the end.

I drove the car back to Johannesburg, turned it in, and was taken to the airport for a 707 flight to Rome and a connecting flight to Frankfurt, where a new adventure began, the flight back to the States. Our flight from the States to Frankfurt was on a 707 jet, which was fast and comfortable. Now, at Rhein-Main AFB, I was scheduled to return on a MATS C-57, a four-engine propellered aircraft. While waiting to board our aircraft, we could see another C-57 take off. Only its nose wheel decided not to go with it. Instead, it rolled down the runway, all by itself. We heard that the tower told them to go to their destination and let them worry about it. Nothing they can do here.

So, we boarded our C-57 and flew to Prestwick, Scotland. Let me tell you about our C-57. The seats were backwards, facing the tail of the plane. I guess they were safer that way, but it was a strange feeling flying backwards. We stopped to refuel at Prestwick, Scotland, for our trip across the Atlantic. We visited the terminal for a while and were then called to return to our plane. On our way back, we met a group of people coming towards us and learned that their C-57 had lost an engine over the Atlantic and had to return to Prestwick.

We were a bit apprehensive about flying over the North Atlantic on this C-57, especially in wintertime. That water would be awfully cold. We boarded our plane and took off without incident, heading for Goose Bay, Labrador.

Our flight included several military dependents on board. Sitting in the seat behind me was a mother with a small infant. Unfortunately, this was not a happy baby. It cried almost all the way to Goose Bay. In addition to the crying baby, the props on the plane would get out of sync and cause the aircraft to vibrate about every ten minutes. So, although it was late at night, there would be no sleep for me. Flying backwards, a crying baby, and out-of-sync props. What a flight. And also, flying over the North Atlantic in the winter. Despite all this, we made it to Goose Bay, a very cold place during the winter season.

Then the next wonderful thing happened. We deplaned and got onto an Air Force bus. The door was open, and the inside was like a freezer, no heat and no driver. Why hadn't they warmed the bus for us? It was about 3 am, but still they could have heated the bus. I was wearing a light summer suit, and it was bitterly cold, and my teeth began to chatter uncontrollably. It was the first time that it had ever happened. Eventually, we got a driver and were driven to the terminal and warmth. Goose Bay is damn cold in the winter.

My next stop was McGuire Air Force Base in New Jersey, and a connecting flight directly to Albuquerque and home. Well, maybe not. Seems I missed that flight. The only flight available to me was to Kansas City with a connecting flight the next day. Another day.

I arrived in Kansas City that night and was taken to a hotel. After checking in, a bellhop took me to my room. As he was leaving, he said, "I have a nice little blond for you waiting downstairs." Surprised, I replied, "No, that won't be necessary, I get home tomorrow." And I did.

Piece Number 10
Our Friends, the Russians

In late 1961, during the Kennedy administration, Russia decided to blockade Berlin again. As a result, President Kennedy canceled all discharges from all the services. With my discharge approaching in February, it would get me for three months and thirteen days. The blockade was eventually resolved, but along came another problem with the Russians. This one did involve me more.

Russia and the United States had signed a treaty in which both parties agreed not to test nuclear weapons in the atmosphere. One day, someone entered our shop and announced that the Russians had broken the treaty and resumed testing in the atmosphere. Where I worked, in telemetry, we were involved in weapons testing, as was most of the Air Force Special Weapons Center. Unfortunately, the United States was not at all prepared to resume testing. The Russians got us on this one.

The word came down to us directly from President Kennedy that the U.S. would resume testing in ninety days. This meant working night and day if necessary. My shop went into three eight-hour shifts per day.

I was assigned to a project that involved determining the exact burst altitude of the bomb using an electronic triangular system. The electronic engineers were the ones under the gun. They had to design a great deal of the circuitry for use in the tests, and they didn't have much time.

The drop plane was a B-52, followed by three C-130s collecting test data. In addition to my primary job, I was a backup for the team in one of the C-130s. So, I got placed on flight status, which meant a little more money in my paycheck. I had the opportunity to participate in a test flight over the Nevada desert. Then it was on to Hawaii, where I would spend a month at Barbers Point Naval Air Station. The testing was done near Christmas Island. Near the end of the testing, I was instructed to return to Kirtland. My discharge had finally been approved and was waiting for me. The next day, I spent most of the day at Hickam AFB waiting for a flight back to the mainland. Then, because I had such a heavy toolbox with me, I had to take a train from San Francisco to Albuquerque. When I got back, I cleared the base and walked out the gate as a civilian. After four years, three months, and thirteen days in the Air Force and three years at Kirtland AFB, there was a bit of sadness in thinking that I would not return or see my Air Force buddies again.

Piece Number 11
Next, the University of New Mexico

One of the civilians who worked in the Telemetry shop was studying electrical engineering at the University of New Mexico, not too far from Kirtland base. He would show me his math problems, and we would discuss them, and I became very interested in math. So, I decided to go back to school and get a degree in mathematics when I was discharged.

Having matured a great deal since my earlier college work, I dedicated my life to mathematics. And I got my degree by working very hard. I'm proud of my 3.5 grade point average in those two and a half years at the University of New Mexico. And again, it was due to very hard work and dedication. My minor was in Electrical Engineering, another difficult subject. I graduated from university in 1965.

During the three years at Kirtland and the two and a half years at UNM, I had complete support from Glenda. She worked during those five and a half years. When I graduated, the University Dames organization at UNM presented her with a "PHT" degree. "Putting Hubby Through". She certainly deserved that and much more.

And now what? Well, I guess I ought to find a job to support her and myself and whatever family there was to follow. It was recruiting time at the university and time for interviews. I had a couple of offers to be a math aide, also one for a statistical job. Then came a recruiter from Phillips Petroleum Company, and he was looking for math majors to train to be computer programmers.

At university, I took a short course in Fortran programming and found it fascinating. That was the job I wanted.

The recruiter for Phillips was the best I had talked to. He was really a nice guy. He offered me a visit to Phillip's facilities in Bartlesville, Oklahoma, which I grabbed.

They flew me to Bartlesville, showed me all their computing equipment and programming facilities then showed me all around Bartlesville. I was very impressed. After a series of interviews, they offered me a job. I delayed accepting it until I talked with Glenda. It sounded great to her, so I accepted the job, and we headed for Oklahoma.

Let me mention a situation like the one I had in the Air Force. I was trained on a piece of equipment that I never saw again, once I got into the field. The same thing happened in my math studies. In both math and electrical engineering, solving differential equations is a must. I must have solved hundreds of them.

Programmers at Phillips don't need to solve differential equations. Never saw another differential equation again. It's the logical thinking of math majors and math teachers that makes a good programmer.

Shortly after hiring on, I received word that my father had died from a heart attack. I returned to San Francisco for the funeral. I felt no hostility toward him, only sorrow over the life we had together. It could have and should have been so much better. I was saddened by my father's death.

Now we were expecting our first child. But there were complications, and we lost her. A time of intense sorrow. A year or so later, however, we had a healthy baby boy, my son Eric.

Before becoming a father, I accepted the belief that my parents suffered from a disease, alcoholism. They couldn't help themselves. But when I became a father and realized how precious my children were, and how I could never let any harm come to them, I was less inclined to forgive my parents. How could you become oblivious to your children and completely abandon them? Will I ever be able to completely forgive you? In 1970, my mother passed away. She had been fired from her manager's job and, of course, began drinking. And it finally killed her—cirrhosis of the liver. 1937 to 1970, the year I was born to the year my mother died, thirty-three years mostly filled with drinking and drunkenness. In 1970, I became free from the alcoholism and drunkenness that destroyed most of my childhood. There would be no more. It was over, finally over. And my new life began, but there was still a great deal of pain ahead.

The portrayal of my mother and father in this book is quite harsh. When drinking or drunk, they deserved it. But there were good times together, times when they were good, lovable parents, times when they were sober. Without their drinking, our lives could have been so different, filled with so much more happiness. Thinking of what could have been saddens me so.

Piece Number 12
My Brother Bill

Many times in this story, I've tried to remember where Bill was and what he was doing during the alcoholism events. But I couldn't. Did I neglect him? I don't know. When I joined the Air Force, I did abandon him, and I have always regretted that. What could I have done instead?

Bill was a good guy and a good father to Karen. He brought his wife, Linda, into our lives, and we were happy for that. Bill and Linda were eventually divorced, and June came into his life. Except for Karen, they are all gone now.

Bill was once written up in his local newspaper because he spent so much time playing ball with the neighborhood kids.

Let me tell you a story about him. He had been in the Army for some time and had to do a tour in Vietnam, which he did, and returned on leave. Later, he told my mother (this is a sober story) that he had to go back to Vietnam. My mother was outraged that they would force him to go back. So, she got on the phone and called the Pentagon in Washington, D.C. and raised holy hell with the Army. Someone called her back later and told her that Bill had volunteered to return to Vietnam. Lying to my mother was not a smart thing to do. I'm sure he caught Hell from the army also.

The good life can still have its rough spots, some being extremely rough— as it was for my brother Bill.

Bill's daughter, Karen, was to be married on a large boat in the middle of San Francisco Bay. Glenda and I flew to San Francisco for the occasion.

And it was a wonderful event and one new to us. We returned home the next day. That same day, Bill was scheduled for some minor surgery to clear a vein in his neck. The following day, a blood clot broke loose and went to his brain. Disaster!

The resulting stroke was terrible. He was bedridden, could not talk, could not swallow food, and several other things he could not do. He had been divorced from Linda and now had a lady friend, June, staying with him.

One morning, June failed to show up at his side for quite a long time. Somehow, Bill got out of bed and crawled into her bedroom, and found her dead. He managed to crawl to the telephone and called the fire department. But he couldn't speak clearly. Fortunately, the person at the station recognized Bill's number and sent help.

Since Bill could not swallow properly, they attached a feeding tube to his stomach. Bill had only one functioning kidney, and it was struggling. From what we were told, the incision for the feeding tube started bleeding. Eventually, the blood flow overwhelmed one kidney, and Bill died.

How I miss that guy, I loved him so. The second part of my life, the part from Glenda on, did have its times of sorrow.

Piece Number 13
A Girl for Us and My Job

We decided to adopt a girl to replace Lorie Lynn. We completed all the paperwork and the background checks required by the state of Oklahoma. And one day, we got a call from Oklahoma City telling us that they had a baby girl for us. We wasted no time getting to the Oklahoma Capitol Building to get our little girl. And we did. Only a few months old, she was just what we wanted. So, we added Diane Cheryl to our family. A good job in a nice town, a wonderful wife, and two wonderful children. Better than winning a million-dollar lottery. The good life followed. Of course, there was sadness at the loss of Glenda's parents and other family members.

We lived in Bartlesville for over thirty years, with twenty-three of those years as a computer programmer for Phillips Petroleum Company.

Early 1965 was not much after the dawn of the computer age. Programs were written mostly in Fortran, and the instructions were punched on what were called IBM cards. I would write my program instructions on a lined paper pad. When finished, I would send my instruction sheet to the key punch operators who punched the holes in the cards. Then the deck of cards went to the machine room. There, it was stacked with many other programs in a card reader. The output from the program was then punched on cards and taken to the card reader of the printer, read in to finally produce a printout.

If a program had a problem and terminated, the computer would produce a core dump. A printout of every cell in the computer in octal rather than decimal. Octal is a numbering system based on the number eight rather than the number ten. After a few years, I became an expert in core dumps.

The laptop computer that I am writing this story on is maybe a thousand times faster and has many, many times as much memory and storage.

My programming work involved seismic exploration for oil and gas. Seismic data was obtained by vibrating the local land to send energy waves downward through the earth. Reflections would occur due to the changes in density in the different layers and would be recorded. Processing on the computer would clarify the layer changes. Oil and gas are accumulated at certain types of structures.

Seeing those structures suggested where to drill. I and many other programmers wrote programs to process this seismic data. At the end of my twenty-three-year career, I was involved in geology, being responsible for a large sub-surface mapping program. The time came again when Phillips needed to downsize. The fourth time offered a very good retirement option for those fifty and older. Having just turned fifty, I grabbed the offer and became unemployed. And it turned out to be one of my best decisions. I retired as a Senior Computing Research Analyst.

So, what does a retired programmer do? Well, he writes programs, or he tries to teach non-technical adults how to program at a vocational tech school. I got through it, but I would never do it again. A third of the class thought I was going too fast, a third thought I was going too slow, and a third didn't care. Same proportions for interest in programming.

At this time, shareware was popular. Shareware was a "try before you buy" option, and there were many programs being sold that way. At the time, the only travel planning program was all wordy and not graphical. It occurred to me that a graphical travel planning program was needed. And I thought that if I did it, I could sell it to AAA and make a lot of money. So, I wrote a graphical travel planner called "The Interstate Traveler". Graphical means there is a map of the United States with interstates, roads, states, cities, and more. You could plan a trip by clicking on your choices with the mouse.

I worked on that program for a year. About a month before finishing it, AAA came out with its own version of the program. They beat me to it, and so did several other companies. My dream went up in smoke, at least the part about making a lot of money. But I did have one big advantage, cost. The other programs required several programmers who got paid salaries. And I had just me. Their programs sold for about $100. I sold mine for $35. I did sell some. It was for sale in Walmart for a while.

Then Windows came out, and competitors jumped on it. I didn't, and that was the end of me and the "Interstate Traveler".

So, what does a retired programmer do now? Write another program, of course. A series of them.

The first was "Fisherman's Delite", followed by "Fisherman's Warmwater Challenge", "Fisherman's Coldwater Challenge", and "Fisherman's Saltwater Challenge". These were more successful by quite a bit.

One day, I started on a new program. After some time, I decided that I could not do it anymore. Enough programming.

I had a business called "JacobSoft". I was CEO, president, vice president, treasurer, secretary, gofer, and janitor.

But it was fun selling my software as well as the software of other programmers. I was in business for ten years, which included several years assembling computers at a Bartlesville computer store. I learned a lot about the insides of a computer.

David Jacobson

Piece Number 14
RVing and Other Travels

Let me back up a few years. While still working, I decided to get into RVing. So, I bought a pickup truck with a camper in the back. It turned out that the camper was crowded with two small kids and Glenda, and me. So, we moved up to the first of a couple of fifth wheels. And we did a lot of camping. After retiring, I decided to trade in the fifth wheel for a camper van. The kids were grown and on their own, so Glenda and I had the Roadtrek van to ourselves. This was about the same time that I quit programming. The Roadtrek was the perfect vehicle for two people. It had everything, a double bed, stove and microwave, refrigerator, enclosed shower, and dining table seats which convert to another double bed. And this started the RVing era for us. We went to every National Park in the western and central U.S., plus the Everglades. We never got to the northeast. This was a very enjoyable time in our lives. But it would get even better.

During this time, we visited New York City twice by air. The first time was in June of 2011. While there, we went to the top of the World Trade Center. Spectacular view of New York City from that height. Taxicabs looked like little bugs crawling around far below. Then came October 11th. That morning, I was sitting on my exercise bike watching TV in front of me when something unbelievable happened. First, I heard that a jetliner had crashed into one of the towers of the World Trade Center. Then, on live TV, I saw the second jetliner go into the second tower. I couldn't believe my eyes. And what followed was horrendous. We returned to New York City a couple of years later, when the foundation of the new building was being prepared.

On the first visit, New York City was a great place to tour. Glenda and I would walk all over the city, night and day, and never felt any danger as you would today in 2025.

We went to a couple of Broadway plays, "Phantom of the Opera" and "The Lion King". Wonderful!

If I lived in New York City, I would go to a Broadway show every week. During our second trip there, we attended "Mamma Mia" and "Thorough Modern Millie". I thought that "Millie" was the best play I'd seen. We loved Broadway and New York City back then, but today in 2025, no love.

We did a number of cruises to the Caribbean. But the best cruise was to Alaska. And the highlight of that trip was a helicopter flight out to a glacier where we were allowed to walk on it. The view from the helicopter was breathtaking, it was over snow-covered mountains, mountain lakes, and the beautiful glaciers. We visited Anchorage and Fairbanks. A wonderful place to be in the summertime.

We visited many special places in our years of RVing—South Dakota, Mount Rushmore, the beautiful coast of Oregon where we found good Tillamook Cheese, Road to the Sun in Montana which I don't think I would go up or down again, Painted Desert, Crater Lake, Death Valley, Utah National Parks, Site of the Little Big Horn, Grand Canyon, Yosemite, Yellowstone. So much to see and do. I am thankful for the opportunity to have seen as much as we did.

We had gone to Arizona to visit several tourist sites. On the way back to Oklahoma, we decided to stop and see some friends living in Sun City, in Georgetown, Texas, about twenty miles north of Austin. We did and enjoyed the visit. Then they took us to see the model homes. And one of those was the perfect house, I loved it. "Glenda, we are moving to

Texas".We talked to the contractor, and he was more than happy to build one for us. He did, and we became happy Texans. It was a wonderful house, and we made a lot of friends. Glenda would agree that this was the best place we ever lived, and that is still true today. We lived there for eight years.

The area beyond the back of our property line was tall grass. We had a doe drop a fawn in that tall grass not far from the house. Mama would be gone all day, and the little one peeked over the grass a few times. With Mama gone, I was worried that something might come along and harm it. But Mama came back every night, and all was well. The two things we had to be concerned about were big feral hogs and rattlesnakes. We were warned not to stick our hands into the bushes. We hired someone to maintain the lawn and bushes. The feral hogs kept chewing up the greens of the golf course, so a hunter was hired to shoot them down.

The best eight years of my life. We made so many friends, did so many activities, and enjoyed so many facilities. Sun City set us up on a cruise and took us to Galveston to board. Glenda and I volunteered to be greeters in a small station at the main entrance. We talked to people from all over the country and enjoyed telling them about Sun City. There were so many activities. We enjoyed playing Bocce ball quite a bit. There was a fishing pond from which I caught a few fish. I did Art and stained glass, and Glenda did quilting for at-risk children with the Linus Club. I wish I could have spent my life there. Then I found one activity that turned into the most enjoyable activity of my life. There were several types of dancing classes. Our friends, the ones we came to visit, were in Line Dancing. And they tried to talk me into trying it. Not being a dancer made me very hesitant to get on the dance floor. I got steps to some of the dances they were doing, practiced often, and gave it a try. Well, it turned out to be one of the best things I've ever done. I not only became a line dancer, but I also became a very good dancer. Newcomers would dance next to me to watch my steps, and most of them were women. So, I got the reputation of having groupies. I line danced for six years, loved every minute of it. Then my legs decided to put a stop to it. Damn! Never again would I do that which I loved and enjoyed so much, line dancing.

Many years previously, I had a serious and very painful problem with my legs that prevented me from walking. It was diagnosed as spinal stenosis, and I needed surgery to fix it. The surgery went well, and my legs were as good as new. I was told that in ten years or so, the problem could return.

While at Sun City, the pain in my legs returned and was severe. I went to my primary care doctor. He told me there was a new orthopedic surgeon in the hospital, and he could fix me up. So, I went to him, and after several tests, he told me I needed a spinal fusion. So, I agreed, assuming he knew what he was doing. But I was wrong. Not long after the surgery, the pain in my legs returned. I suspect that my fusion was the first one this surgeon had done. What I didn't do was going to cost me a lot. Glenda had a spinal problem earlier and had it fixed by a neurosurgeon whom we liked very much. So why didn't I get a second opinion from him?

If only I had, I probably wouldn't be in a wheelchair today and for the last ten years. Let me stress a point: "GET A SECOND OPINION." I didn't, and I'm paying for it.

When I returned to my primary care doctor and told him that the pain had returned, he apologized for sending me to that surgeon. We decided to return to Oklahoma, where our son and daughter still lived. There was no problem selling the house, except for the unhappiness of having to leave it. And we returned to Oklahoma City. There, I went to a neurosurgical group, where they did an MRI of my spine. When the senior neurosurgeon saw it, he became very angry. "Why didn't they finish this?" It took two neurosurgeons six hours to remove the hardware in my back and replace it properly. The first failed surgery fused five vertebrae, and the second fused six more vertebrae, eleven in all. I now have two rods and twenty-two screws in my spine. They told me there was quite a bit of nerve damage. That was ten years ago, and it's been downhill since then. That 16-inch scar on my back, which has been opened three times, marked a major change in my life. The pain in my lower back has continued for ten years. And I have developed intense pain in the backside of both thighs. Because of this, I can't sit in a chair, especially in my wheelchair, for very long. When the pain gets bad, I must lie down in bed. This happens five or six times a day. And as the years have gone by, my legs have gotten weaker.

I've fallen six times in the last 6 months but haven't broken anything yet. I can't climb a single stair or get into a car without help. At my age, I won't let them cut me up anymore. So, I'm taking many pills and capsules plus patches, none of which are doing any good.

For the last 5 years, we have lived in several independent living facilities. The last couple of years, it's been assisted living.

About fifty years ago, I developed restless leg syndrome. I've gone through all the stages and taken so much dopamine that it doesn't help anymore. And there are no alternatives. It's a terrible syndrome mainly because it frequently prevents me from getting any sleep. It can be maddening at times. Even when asleep, my legs can be kicking, keeping Glenda awake.

My health was good during my fifties and sixties, but the seventies and eighties have not been very good. Some advice: Do the things you want to do, cruises, travel, hobbies, or learn something new while you can. Once your health deteriorates, those things can become meaningless.

Putting the Pieces Together

I have continued my artwork through all these years. Art has been my number one hobby. I started with oils, then acrylics, watercolors, and finally digital art on the computer. I've spent many hours at the keyboard and completed hundreds of digital art pieces. The best thing about digital is that you can change anything anytime. If you don't like the color or shape, just change it. That is difficult to do with paints. I also like watercolors with marker or pen. I do wish I could have more time for my art, though it is getting harder to think up new subjects. My favorite subject is music. I have done many landscapes, skyscapes, seascapes, abstracts, figures, Oklahoma, Kokopelli, patriotic, military, police, and fire, and something I call crazyscapes. I've scattered some of my digital art throughout the book, not necessarily relevant to my story. I've always tried to develop something new and different, but I'm still trying. I have a new style that I'd like to try out when I finish this book.

Piece Number 15
The Last Piece

Now a word or two about my son, Eric, and my daughter Diane.

Eric did four years in the Marines, and I salute him and all Marines for the great, but difficult, job they do. He followed the Marines with thirty years in the Oklahoma City Police Force, another difficult job. He has done his share. I salute all police, fire, and emergency medical personnel. We wish Eric a wonderful retirement! And our love to Krista, his wonderful wife.

My daughter, Diane, a college graduate, has been a single mother for more than fifteen years. She works hard, but still makes my grandson, Angus, the most important part of her life. And she is always there when we need her. We are blessed with two wonderful children.

Let me say with pride that this family has served its country: Father, Army, Philippines; brother Frank, Army, WWII; brother Bill, Army, Vietnam; son Eric, Marines, Panama; and me, Air Force.

For the last few years, it has been independent living and now, assisted living facilities. When the time comes, it comes.

The most important takeaway from my story is that it is possible to overcome adversity, to survive, and to prosper. Some of my days were so dark that I thought I'd never make it to adulthood. But I did, and I have prospered. You can too, if necessary. Work hard at whatever you do and be a good person through it all, and don't give up hope. Ask for help when you need it. Contribute to the goodness in the world and cast down the evil in it.

Putting the Pieces Together

We all have stories to tell, many more adventurous and some more tragic than mine, but each of us has something important to say. For me, this book puts together all the pieces of my life. Maybe we all should put the pieces together to see where we came from and where we are, and where we are going.

The opinions in this segment are mine and mine alone. I am responsible for my own thoughts. First, some thoughts about my hometown.

Back in the 1960s and later, San Francisco was a good place to visit, as we did several times. The Chinese food was special, the pizzas were great, and a shrimp cocktail with a basket of French bread at Fisherman's Wharf was my favorite. I must not forget my favorite vegetable, artichokes. When I lived there, it was a great place to visit and live in, and to enjoy many interesting sites, Golden Gate Park, Golden Gate Bridge, Chinatown, Twin Peaks, Fisherman's Wharf, Palace of Fine Arts, Civic Center, Alcatraz, Bay Bridge, Coit Tower, and more. I haven't been there in years. I've seen some videos of the City, and it's not the city I had lived in. Empty buildings, trash, and homeless people everywhere, traffic jams, crime, and more. I don't think I'd want to go back there. And the possible "Big One" would have me concerned if I lived there. Speaking of earthquakes, I have experienced several of them when I lived there. This one time, I was sitting at a desk, sitting on a chair with wheels, and my feet were off the floor on the chair. All of a sudden, I was rolling across the floor and ran into the wall behind me. That was a strange sensation.

I write this book at the dawn of the "AI" or "Artificial Intelligence" age. I can't imagine what computers, cars, aircraft, ships, and toasters will look like in 5 years. Will robots rule the earth in 10 years?

Mother Earth or Father Universe may decide to give humanity a good spanking. Today in 2025, we seem to be plagued with earthquakes and volcanic eruptions, plus the threats from flares from the sun. There is talk of large space objects hitting the Earth or the Moon. Plus, the possibility of alien life being detected. I hope to live long enough to experience the next five or so years. The possibilities are staggering, as are the threats. And today, civil or world wars are threats to our country. The hatred within our country threatens our existence. My look at the future right now is quite negative. Too much hatred, both at home and around the world. Hate begets hate. Why can't people understand that? Maybe God will step in and save humanity, if humanity is worth saving. Five years from now, we may all be gone by then. Think about it! I hope to be whisked away before the world collapses.

If AI puts too many people out of work, they will demand that the government take care of them. Sounds like Socialism, a step towards Communism. "People's Republic of the United States"? Gloom and doom, that's me right now. Let us demand the disappearance of HATE! AI will be something that the world has never experienced. What will it bring? It is your future, what will it bring you?

I wish you, my friend, good fortune. Before ending this book, I must mention five members of our family who we loved dearly, Muffin, Cindy, Cassie, Reggie, and Toby. What joy they brought us.

Thank you for reading my story.

www.ingramcontent.com/pod-product-compliance
Lightning Source LLC
Chambersburg PA
CBHW051240120626
46547CB00014B/1726